I0469129

U.S. Fire Administration
Mission Statement

As an entity of the Federal Emergency Management Agency, the mission of the United States Fire Administration is to reduce life and economic losses due to fire and related emergencies, through leadership, advocacy, coordination, and support. We serve the Nation independently, in coordination with other Federal agencies, and in partnership with fire protection and emergency service communities. With a commitment to excellence, we provide public education, training, technology, and data initiatives.

On March 1, 2003, FEMA became part of the U.S. Department of Homeland Security. FEMA's continuing mission within the new department is to lead the effort to prepare the Nation for all hazards and effectively manage Federal response and recovery efforts following any national incident. FEMA also initiates proactive mitigation activities, trains first responders, and manages Citizen Corps, the National Flood Insurance Program, and the U.S. Fire Administration.

FEMA

United States Fire Administration
16825 South Seton Avenue
Emmitsburg, MD 21727

A Fire Safety Campaign for People 50-Plus

Prevent Fire.
Save Lives.

Dear Fire Service Member,

A Fire Safety Campaign for People 50-Plus is a new campaign developed by the United States Fire Administration for our Fire Service partners. We are writing to ask you and your colleagues to join us in this fire safety education effort targeted to people ages 50-plus, their families and caregivers.

Each year, approximately 1,100 Americans ages 65 and older die in home fires and another 3,000 are injured. People ages 65 and older are three times as likely to die in a residential fire as the rest of the population. These statistics, combined with the fact that people ages 50 and more care for and will soon enter this high-risk group, inspired this campaign from USFA, a division of the Federal Emergency Management Agency (FEMA) and part of the Department of Homeland Security.

The campaign encourages people ages 50-plus to practice fire safe behaviors when smoking, cooking, and heating their homes. It also stresses the importance of maintaining smoke alarms, designing and practicing a fire escape plan, and if at all possible, installing home fire sprinklers.

In this *Campaign Guide* you will find:

- A "how to" introduction to using materials in the *Campaign Guide*, and to using the computer CD with all of the campaign materials.
- Campaign fact sheet and flier – in English and Spanish – that can easily be copied and distributed in your community.
- Sample media materials, including a news release and live-read radio script, a print public service ad, and tips on how to get the media to use your fire safety materials.

In addition, these free campaign materials are all available on the campaign's Web site, www.usfa.fema.gov/50Plus.

If you have requests or questions, please use the enclosed feedback form or e-mail usfa@hagersharp.com. Thank you for your assistance with *A Fire Safety Campaign for People 50-Plus*. Working together, we can reduce fire fatalities among our high-risk older population.

Sincerely,

Kathy Gerstner

Kathy Gerstner
National Fire Programs
U. S. Fire Administration

 FEMA   www.usfa.fema.gov/50Plus

Campaign Guide

A Fire Safety Campaign for People 50-Plus

Contents

Campaign Overview

The Statistics We Need to Change

- People between 65 and 74 are nearly twice as likely to die in a home fire as the rest of the population.
- People between 75 and 84 are nearly four times as likely to die in a fire.
- People ages 85 and older are more than five times as likely to die in a fire.

Each year, approximately 1,100 Americans 65 and older die in home fires and another 3,000 are injured. These statistics, combined with the fact that adults ages 50 or more care for and will soon enter this high-risk group, inspired USFA to develop a new public education campaign targeting people ages 50-plus, their families and caregivers.

The People 50-Plus Campaign Strategy

A Fire Safety Campaign for People 50-Plus encourages people ages 50 and older – including the high risk 65-plus group – to practice fire-safe behaviors to reduce fire deaths and injuries. The strategy is to inform and motivate adults as they enter their fifties so that stronger fire safety and prevention practices are integrated into their lives prior to entering the higher fire-risk decades. In addition, many Baby Boomers are currently caring for family members ages 65-plus and can encourage fire safe habits.

Messages That Will Make A Difference

As part of the Fire Service, you know firsthand that older people most often suffer fire death and injury because of careless smoking, cooking or heating. These are the messages that we must drive home, beginning with people ages 50-plus:

- **Use Smoking Materials Safely:** Don't smoke in bed, while drowsy from medications or alcohol, or when you're just plain sleepy. Use large, deep ashtrays for smoking debris, and put your cigarette or cigar all the way out. Don't walk away from a lit cigarette.

- **Pay Attention to Your Cooking:** Turn pot handles inward, and keep cooking surfaces clean and clutter-free. Don't wear loose clothing while cooking and never leave cooking unattended. Double-check the kitchen before you go to bed or leave the house.

- **Heat Your Home Safely:** Keep flammable materials at least three feet from heaters. When buying a space heater, look for a control feature that automatically shuts off the power if the heater falls over. Make sure you have a fireplace screen large enough to catch flying sparks and rolling logs.

- **Install and Maintain Smoke Alarms:** Install a smoke alarm on every level of your home, test batteries every month and change them at least once a year.

- **Home Fire Escape Plan:** Develop and practice a fire escape plan regularly, at least twice a year. Keep exits clear of debris.

- **Install Home Fire Sprinklers:** If at all possible, install residential fire sprinklers in your home.

Getting the Messages to People 50-Plus in Your Community

This is *your* campaign – and USFA has developed it with you in mind.

- Materials in English and Spanish are high quality black & white – easy and inexpensive to copy from the *Campaign Guide*, right away.

- A computer CD with campaign materials makes it easy to reproduce materials at a copy shop or printer.

- **www.usfa.fema.gov/50Plus** – the campaign Web site – is a ready source of fire safety information for people 50-plus in your response area.

Read on for tips on community and local media outreach – and to see campaign materials and sample media materials. USFA will provide momentum to your efforts by engaging fire departments and fire marshals nationwide, and by working with national media to promote campaign messages.

How to Use the Campaign Materials

All materials of *A Fire Safety Campaign for People 50-Plus* are designed to help with your public education outreach. Using the *Campaign Guide*, you can plan your department's own *People 50-Plus* campaign in several easy steps.

Using the *Campaign Guide*

1. **Plan your *People 50-Plus* outreach** using tips from "Working in the Community."

   Make copies of the free materials from the *Campaign Guide* – in English and Spanish – for handouts. Easy-to-copy black & white materials include:

   - The 8 ½" x 11" **flier**.
   - The two-page **fact sheet**.
   - **Bookmarks.** Copy onto heavier paper and cut down for great handouts.

2. **Engage your local media** in your department's *People 50-Plus* campaign. Read "Working With the Media" for ideas on how to adapt and use media outreach materials in the *Campaign Guide*, including:

   - **News Release** announcing your department's People 50-Plus Campaign. Add your department's name and quote your chief. Adapt it for your community.
   - **Print PSA** (public service advertisement) for your local newspapers. See "Using the Computer CD" for instructions on how to send the paper a camera-ready ad.
   - **Script** for radio and television stations. Radio announcers can use it as a "live read script" – or get a fire department spokesperson to record it. Encourage your local TV stations to use it for on-air "tips" when they have a few extra seconds.

Using the Computer CD

The computer CD accompanying your *Campaign Guide* contains **all materials in the guide** plus a **poster**, **PowerPoint slides** for talks and displays in your community, and a **data report** with statistics on fire fatalities among people 50 - 64 (national data) and people 65-plus (national and state-by-state data). Use the computer CD to:

1. Print copies of materials from your computer.
2. Take it to a library, copy shop or commercial printer to makes copies of materials.
3. Once your newspaper has agreed to run the print PSA, send the paper an electronic version of the ad – or lend them the CD to make a copy.
4. E-mail materials to firefighters and community groups throughout your response area.

Using the Campaign Web Site

www.usfa.fema.gov/50Plus has all of the campaign materials in large typeface for easy use by people 50-plus in your response area. It's also a quick way to share campaign information with the media and firefighters throughout your community. The site is another source for you to read and print campaign materials.

Working in the Community

Community outreach is a regular part of most firefighters' lives. The following list will help you focus outreach efforts on people ages 50 and older. It includes organizations to contact for help in identifying chapters in your area, setting up group presentations, and including fire safety information in home visits with people ages 50-plus. Many of the organizations hold regular meetings and would provide an opportunity for you to teach fire safety.

If you plan a group presentation, be sure to copy materials from the *Campaign Guide* or computer CD to distribute. When equipment is available, you may want to supplement your talk with PowerPoint slides on the computer CD.

For home visits, point out potential fire hazards in your hosts' homes and help them to draft a fire escape plan. Remember to bring plenty of materials to leave behind – and tell 50-plus Internet users about **www.usfa.fema. gov/50Plus**. When Meals on Wheels workers and visiting nurses join your *People 50-Plus* campaign, be sure they have plenty of leave-behind materials, too.

- **AARP**: Call 1-888-687-2277 for contact information for your local chapters.
- **Area Agencies on Aging (AAA)**: Call the *Eldercare Locator* at 1-800-677-1116 for local divisions or visit **www.eldercare.gov**. The AAA coordinates many services for older adults.
- **Meals on Wheels Association of America**: Call 703-548-5558 or visit www.mowaa.org for your state contacts.
- **Retirement and Nursing Homes**: Plan a visit to your local facilities and bring enough materials to leave behind for the residents.

- **Senior Centers**: Ask to schedule a meeting and bring extra materials.
- **Faith-Based Organizations**: Contact your local churches, mosques and synagogues. Request the opportunity to present fire safety materials at regular sessions for members ages 50-plus. Conduct similar outreach through faith-based groups like the Knights of Columbus.
- **Visiting Nurse Associations of America**: Call 617-737-3200 or visit **www.vnaa.org** to identify local divisions. Provide these healthcare professionals with materials to distribute on home visits.
- **Local Library**: Many libraries have community areas where you can post the fact sheet or leave copies of the fliers with tips.
- **Community Center**: Many community centers have community boards for posting events or information. Post the flier or fact sheet on the boards and leave copies so that residents can take them home.
- **Grocery Stores**: Contact local corporate offices of grocery chains and independent stores. Many are quite community-oriented, and will welcome your request to post the campaign flier and fact sheet on consumer bulletin boards.

Working with the Media

Your local media are great vehicles for reaching a larger audience and relaying fire safety messages for people ages 50 and older. The following tips will help you get results from the media. The *Campaign Guide* also includes sample media materials that you can adapt for your community and put on your fire department's letterhead.

Get Media Results from Materials and Spokesperson Interviews

- Send the news release announcing your department's new *Fire Safety Campaign for People 50-Plus* (see sample in the following pages) to all newspapers, radio and television stations in your area.
- Call local TV and radio stations and speak with the assignment editor (TV) or news director (radio). Alert them to the campaign. Offer to fax or e-mail them the campaign fact sheet – and send the news release again if necessary.
- Make similar calls to the newspaper editors and reporters who cover "fire," "personal health and safety" and "seniors" beats and offer to fax or deliver materials in person.
- Designate a campaign spokesperson and offer the media interviews with him or her.
- Facts are what make a news story. Review the campaign's messages (see "Campaign Overview") and use the statistics about people 50-plus and fire risk. You can also get statistics specific to your state from the "data report" on the computer CD or www.usfa.fema.gov/50Plus.
- Talk about a recent local fire in the interview. An anecdote that people can relate to makes the story more powerful, and relays the importance of fire safety.
- Have your spokesperson practice the messages before conducting the interview.

Ask Radio Stations to Air the Campaign Public Service Announcement (PSA)

- Adapt the sample radio script in the *Campaign Guide* to your department.
- Send the script to public service directors at your local radio stations along with a cover letter (sample included in *Campaign Guide*) on your department's letterhead. The letter describes the campaign and asks for the radio station's support.
- Call the public service directors to confirm that they received the PSA script and tell them briefly why this campaign is so important. Follow-up can make all the difference!
- Offer to provide a fire department spokesperson to record the radio PSA. The station may prefer for an announcer to use it as a "live-read" script.

Ask Local Newspapers and Magazines to Run the Print Public Service Ad (PSA)

- Call the newspapers and magazines and ask for a contact person for public service ads.
- Send a cover letter and a copy of the print ad (from the *Campaign Guide*) to the contact.
- Call the contact, urge the publication to run the ad, and offer to provide a camera-ready electronic version of the ad from the campaign CD. Either e-mail the print ad to the contact, direct them to the ad on www.usfa.fema.gov/50Plus, or lend them your CD.

Campaign Materials

Print Public Service Ads (PSAs) and Campaign Flier

Print Public Service Ads

The full-page print PSA and the smaller PSAs (turn page) are variations on the signature campaign ad. The ad sizes are designed to fit newspaper, newsletter and magazine requirements. Please provide your local newspaper and other media who agree to run the print PSA with an electronic version of the ads. You can e-mail the electronic file from your campaign CD, meet with media representatives and allow them to copy from your CD, or direct media to the print PSAs in the Media section of www.usfa.fema.gov/50Plus.

Flier

The full-page print PSA (turn page) doubles as a flier. Just copy the full-page version on 8 1/2" X 11" paper and distribute it during your outreach efforts. You can also print the full-page PSA/flier electronically from the campaign computer CD.

She lost her home, too.

It only takes seconds for a lifetime of memories to go up in flames. Unfortunately, as we grow older, our risk of dying in a home fire goes up dramatically. If you are age 50 or older, please pay special attention to fire safety. For your well-being and others you love.

- Don't smoke when you're sleepy—and really put that cigarette out.

- Keep heaters at least three feet from your bed, curtains, or other flammable materials.

- Never wear loose clothing when you're cooking.

- Test your smoke alarms monthly; change the batteries at least once a year.

- Develop and practice a fire escape plan.

- If at all possible, install home fire sprinklers.

Prevent Fire. Save Lives.

 FEMA

To find out more about lowering your risk of fire death and injury, visit **www.usfa.fema.gov/50Plus**.

She lost her home, too.

It only takes seconds for a lifetime of memories to go up in flames. Unfortunately, as we grow older, our risk of dying in a home fire goes up dramatically. If you are age 50 or older, please pay special attention to fire safety. For your well-being and others you love.

- Don't smoke when you're sleepy—and really put that cigarette out.

- Keep heaters at least three feet from your bed, curtains, or other flammable materials.

- Never wear loose clothing when you're cooking.

- Test your smoke alarms monthly; change the batteries at least once a year.

- Develop and practice a fire escape plan.

- If at all possible, install home fire sprinklers.

Prevent Fire. Save Lives.

FEMA

To find out more about lowering your risk of fire death and injury, visit **www.usfa.fema.gov/50Plus**.

She lost her home, too

It only takes seconds for a lifetime of memories to go up in flames. If you are age 50 or older, please pay special attention to fire safety.

- Don't smoke when you're sleepy.

- Keep heaters at least three feet from flammable materials.

- Never wear loose clothing when you're cooking.

- Test your smoke alarm monthly.

- Develop and practice a fire escape plan.

- If possible, install home fire sprinklers.

Prevent Fire. Save Lives.

FEMA

To find out more about lowering your risk of fire death and injury, visit **www.usfa.fema.gov/50Plus**.

She lost her home. too.

It only takes seconds for a lifetime of memories to go up in flames. Unfortunately, as we grow older, our risk of dying in a home fire goes up dramatically. If you are age 50 or older, please pay special attention to fire safety. For your well-being and others you love.

- Don't smoke when you're sleepy—and really put that cigarette out.

- Keep heaters at least three feet from your bed, curtains, or other flammable materials.

- Never wear loose clothing when you're cooking.

- Test your smoke alarms monthly; change the batteries at least once a year.

- Develop and practice a fire escape plan.

- If at all possible, install home fire sprinklers.

Prevent Fire.
Save Lives.

FEMA To find out more about lowering your risk of fire death and injury, visit **www.usfa.fema.gov/50Plus.**

Campaign Bookmarks

The campaign bookmarks are a great way to get the campaign messages out to the public. Print this page of bookmarks on heavy paper stock and cut to size.

She lost her home, too.

It only takes seconds for a lifetime of memories to go up in flames. If you are age 50 or older, please pay special attention to fire safety.

- Don't smoke when you're sleepy.
- Keep heaters at least three feet from flammable materials.
- Never wear loose clothing when you're cooking.
- Test your smoke alarm monthly.
- Develop and practice a fire escape plan.
- If possible, install home fire sprinklers.

Prevent Fire.
Save Lives.

 FEMA

For more information on lowering your risk of fire death and injury, visit www.usfa.fema.gov/50Plus.

She lost her home, too.

It only takes seconds for a lifetime of memories to go up in flames. If you are age 50 or older, please pay special attention to fire safety.

- Don't smoke when you're sleepy.
- Keep heaters at least three feet from flammable materials.
- Never wear loose clothing when you're cooking.
- Test your smoke alarm monthly.
- Develop and practice a fire escape plan.
- If possible, install home fire sprinklers.

Prevent Fire.
Save Lives.

 FEMA

For more information on lowering your risk of fire death and injury, visit www.usfa.fema.gov/50Plus.

She lost her home, too.

It only takes seconds for a lifetime of memories to go up in flames. If you are age 50 or older, please pay special attention to fire safety.

- Don't smoke when you're sleepy.
- Keep heaters at least three feet from flammable materials.
- Never wear loose clothing when you're cooking.
- Test your smoke alarm monthly.
- Develop and practice a fire escape plan.
- If possible, install home fire sprinklers.

Prevent Fire.
Save Lives.

 FEMA

For more information on lowering your risk of fire death and injury, visit www.usfa.fema.gov/50Plus.

United States Fire Administration
16825 South Seton Avenue
Emmitsburg, MD 21727

A Fire Safety Campaign for People 50-Plus

Fire Safety Facts For People 50-Plus

Each year, approximately 1,100 Americans ages 65 and older die as a result of a home fire. Compared to the rest of the U. S. population:

- People between 65 and 74 are nearly TWICE as likely to die in a fire.
- People between 75 and 84 are nearly FOUR times as likely to die in a fire.
- People ages 85 and older are more than FIVE times as likely to die in a fire.

With a few simple steps, older people can dramatically reduce their risk of death and injury from fire. These facts, combined with the knowledge that adults ages 50 and older are entering and caring for this high risk group, inspired the U. S. Fire Administration (USFA), a division of the Federal Emergency Management Agency (FEMA) and part of the U.S. Department of Homeland Security, to develop a national public safety campaign for adults ages 50 and older, their families and caregivers. USFA encourages you to:

Prevent Fire. Save Lives.

For your well-being and others you love:
Practice fire-safe behaviors when smoking, cooking and heating. Maintain smoke alarms, develop and practice a fire escape plan, and if possible, install home fire sprinklers.

Smoke Safely

Sitting in your favorite chair and having a cigarette after dinner seems to some like a great way to relax – but cigarettes and relaxing can be a deadly mix. Falling asleep while smoking can ignite clothing, rugs and other materials used in uphol-stered furniture. Using alcohol and medications that make you sleepy compound this hazard.

Careless smoking is the leading cause of fire deaths and the second leading cause of injuries among people ages 65 and older. Cigarettes when not properly extinguished continue to burn. When a resting cigarette is accidentally knocked over, it can smolder for hours before a flare-up occurs. Before you light your next cigarette, remember:

- Never smoke in bed.
- Put your cigarette or cigar out at the first sign of feeling drowsy while watching television or reading.
- Use deep ashtrays and put your cigarettes all the way out.
- Don't walk away from lit cigarettes and other smoking materials.

www.usfa.fema.gov/50Plus

Cook Safely

Many families gather in the kitchen to spend time together, but it can be one of the most hazardous rooms in the house if you don't practice safe cooking behaviors. **Cooking is the third leading cause of fire deaths and the leading cause of injury among people ages 65 and older.**

It's a recipe for serious injury or even death to wear loose clothing (especially hanging sleeves), walk away from a cooking pot on the stove, or leave flammable materials, such as potholders or paper towels, around the stove. Whether you are cooking the family holiday dinner or a snack for the grandchildren:

- Never leave cooking unattended. A serious fire can start in just seconds.
- Always wear short or tight-fitting sleeves when you cook. Keep towels, pot holders and curtains away from flames.
- Never use the range or oven to heat your home.
- Double-check the kitchen before you go to bed or leave the house.

Heat Your Home Safely

During winter months, December, January and February, there are more home fires than any other time of year. Heating devices like space heaters and wood stoves make homes comfortable, but should be used with extra caution. **Heating is the second leading cause of fire death and the third leading cause of injury to people ages 65 and older.**

Many of these deaths and injuries could be prevented with safe heating practices. So before you grab a good book and cozy up to the fireplace, make sure you do the following:

- Keep fire in the fireplace by making sure you have a screen large enough to catch flying sparks and rolling logs.
- Space heaters need space. Keep flammable materials at least three feet away from heaters.
- When buying a space heater, look for a control feature that automatically shuts off the power if the heater falls over.

The "Get Out Alive" Home Fire Safety Steps

- **Smoke Alarms**: Install a smoke alarm on every level of your home, test batteries every month and change them at least once a year.
- **Home Fire Escape Plan**: Develop and practice a fire escape plan regularly, at least twice a year. Keep exits clear of debris.
- **Home Fire Sprinklers**: If at all possible, install residential sprinklers in your home.

For more fire prevention information, please contact:

Publications Office
United States Fire Administration
16825 South Seton Avenue
Emmitsburg, MD 21727
1-800-561-3356
www.usfa.fema.gov/50Plus

PowerPoint Slides

Prevent Fire. Save Lives.

A Fire Safety Campaign for People 50-Plus

FEMA

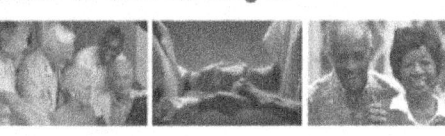

Over 50 and Loving It!

- Retirement
- Grandchildren
- Traveling
- Taking care of older relatives
- New career
- Practicing safe cooking, smoking, and heating behaviors.

Did you know...

- Fire kills approximately 1,100 people ages 65 and older each year.
- People between ages 65 and 74 are TWICE as likely to die in a home fire.
- People between ages 75 and 84 are nearly FOUR times as likely to die in a home fire.
- People ages 85 and older are more than FIVE times as likely to die in a home fire.
- You can do something about it...

Smoke Safely

- Careless smoking is the leading cause of fire deaths among people ages 65 and older.

- Careless smoking is the second leading cause of fire injuries among people ages 65 and older.

Smoke Safely

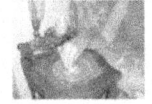

- If alcohol or medication makes you drowsy, or you're just plain tired, put your cigarette out right away.
- Never smoke in bed.
- Use deep ashtrays and put your cigarette all the way out.
- Don't walk away from a lit cigarette.

Cook Safely

- Cooking is the leading cause of fire injuries for people ages 65 and older.

- Cooking is the third leading cause of fire deaths for people ages 65 and older.

Cook Safely

- Never leave cooking unattended. A serious fire can start in just seconds.
- Always wear short or tight-fitting sleeves when you cook. Keep towels, pot holders and curtains away from flames.
- Never use the range or oven to heat your home.
- Double-check the kitchen before you go to bed or leave the house.

Heat Your Home Safely

- Heating is the second leading cause of fire death for people ages 65 and older.

- Fires caused by heating are the third leading cause of fire injury for people ages 65 and older.

Heat Your Home Safely

- Space heaters need space. Keep heaters at least three feet away from your bed, curtains, and flammable materials.
- When buying a space heater, look for a control feature that automatically shuts off the power if the heater falls over.
- Keep fire in the fireplace by making sure you have a screen large enough to catch flying sparks and rolling logs.

Get Out Alive

- **Smoke Alarms**: Install a smoke alarm on every level of your home, test batteries monthly and change them at least once a year.
- **Home Fire Escape Plan**: Develop and practice a fire escape plan regularly, at least twice a year. Keep exits clear of debris.
- **Home Fire Sprinklers**: If at all possible, install residential sprinklers in your home.

For more fire prevention information, please contact:

Publications Office

United States Fire Administration

16825 South Seton Avenue

Emmitsburg, MD 21727

1-800-561-3356

www.usfa.fema.gov/50Plus

Spanish Campaign Materials

Anuncios de Servicio Público (PSA, por su sigla en inglés) Impresos y Folleto de Campaña

Anuncio de Servicio Público Impreso

El anuncio PSA impreso a toda página en ubicación preferencial y los diferentes tamaños de los PSAs en esta página corresponden a los anuncios de la campaña principal. Estos anuncios están diseñados para satisfacer las necesidades de publicación en periódicos, boletines de noticias y revistas. Con toda seguridad, su periódico local o cualquier otro medio de comunicación querrá la versión electrónica del anuncio. Podrá encontrarla en el disco CD o en el sitio de internet de la campaña. Envíe el anuncio en forma electrónica o reúnase con un representante del periódico para permitirle que haga una copia de su disco.

Folleto

El anuncio PSA impreso a toda página en ubicación preferencial también se puede utilizar como folleto. Solamente tiene que copiar la versión a toda página en una hoja de papel de 8 1/2" X 11" y distribuirla entre el público al cual se dirige. También puede imprimir la página completa del PSA/folleto en forma electrónica desde el disco CD.

Ella también perdió su casa.

Toda una vida de recuerdos puede hacerse cenizas en cuestión de segundos. Si usted tiene 50 años o más, preste atención en particular a la seguridad contra incendios. Por su bien y el de sus seres queridos.

- No fume cuando tenga sueño.
- Mantenga los calefactores por lo menos a un metro de las materiales inflamables.

- Nunca use ropa suelta mientras cocina.
- Revise las alarmas de humo todos los meses.
- Desarolle y practique un plan de escape en caso de incendio.

- Si existe la posibilidad, instale rociadores hogareños contra incendio.

Prevent Fire.
Save Lives.

 **FEMA**   Para saber más sobre cómo reducir el riesgo de muerte visite el sitio **www.usfa.fema.gov/50Plus**.

Ella también perdió su casa.

Toda una vida de recuerdos puede hacerse cenizas en cuestión de segundos. Lamentablemente, a medida que envejecemos el riesgo de morir en un incendio hogareño aumenta en forma drástica. Si usted tiene 50 años o más, preste atención en particular a la seguridad contra incendios. Por su bien y el de sus seres queridos.

- No fume cuando tenga sueño y asegúrese de apagar bien el cigarrillo.
- Mantenga los calefactores por lo menos a un metro de la cama, las cortinas y otros materiales inflamables.
- Nunca use ropa suelta mientras cocina.

- Revise las alarmas de humo todos los meses; cambie las baterías por lo menos una vez al año.
- Desarolle y practique un plan de escape en caso de incendio.
- Si existe la posibilidad, instale rociadores hogareños contra incendio.

Prevent Fire.
Save Lives.

FEMA

Para saber más sobre cómo reducir el riesgo de muerte o lesión por incendio, visite el sitio **www.usfa.fema.gov/50Plus**.

Ella también perdió su casa.

Toda una vida de recuerdos puede hacerse cenizas en cuestión de segundos. Si usted tiene 50 años o más, preste atención en particular a la seguridad contra incendios. Por su bien y el de sus seres queridos.

- No fume cuando tenga sueño.
- Mantenga los calefactores por lo menos a un metro de las materiales inflamables.
- Nunca use ropa suelta mientras cocina.
- Revise las alarmas de humo todos los meses.
- Desarolle y practique un plan de escape en caso de incendio.
- Si existe la posibilidad, instale rociadores hogareños contra incendio.

Prevent Fire.
Save Lives.

 FEMA

Para saber más sobre cómo reducir el riesgo de muerte visite el sitio **www.usfa.fema.gov/50Plus**.

Ella también perdió su casa.

Toda una vida de recuerdos puede hacerse cenizas en cuestión de segundos. Lamentablemente, a medida que envejecemos el riesgo de morir en un incendio hogareño aumenta en forma drástica. Si usted tiene 50 años o más, preste atención en particular a la seguridad contra incendios. Por su bien y el de sus seres queridos.

- No fume cuando tenga sueño y asegúrese de apagar bien el cigarrillo.

- Mantenga los calefactores por lo menos a un metro de la cama, las cortinas y otros materiales inflamables.

- Nunca use ropa suelta mientras cocina.

- Revise las alarmas de humo todos los meses; cambie las baterías por lo menos una vez al año.

- Desarolle y practique un plan de escape en caso de incendio.

- Si existe la posibilidad, instale rociadores hogareños contra incendio.

Prevent Fire.
Save Lives.

FEMA

Para saber más sobre cómo reducir el riesgo de muerte o lesión por incendio, visite el sitio **www.usfa.fema.gov/50Plus**.

Marcadores de Libro de la Campaña

Los marcadores de libro de la campaña constituyen una excelente manera de hacer llegar los mensajes de la campaña al público. Imprima esta página de marcadores de libro en papel grueso y recórtelos de acuerdo al tamaño correspondiente.

Ella también perdió su casa.

Toda una vida de recuerdos puede hacerse cenizas en cuestión de segundos. Si usted tiene 50 años o más, preste atención en particular a la seguridad contra incendios. Por su bien y el de sus seres queridos.

- No fume cuando tenga sueño.
- Mantenga los calefactores por lo menos a un metro de las materiales inflamables.
- Nunca use ropa suelta mientras cocina.
- Revise las alarmas de humo todos los meses.
- Desarolle y practique un plan de escape en caso de incendio.
- Si existe la posibilidad, instale rociadores hogareños contra incendio.

Para saber más sobre cómo reducir el riesgo de muerte visite el sitio www.usfa.fema.gov/50Plus.

Ella también perdió su casa.

Toda una vida de recuerdos puede hacerse cenizas en cuestión de segundos. Si usted tiene 50 años o más, preste atención en particular a la seguridad contra incendios. Por su bien y el de sus seres queridos.

- No fume cuando tenga sueño.
- Mantenga los calefactores por lo menos a un metro de las materiales inflamables.
- Nunca use ropa suelta mientras cocina.
- Revise las alarmas de humo todos los meses.
- Desarolle y practique un plan de escape en caso de incendio.
- Si existe la posibilidad, instale rociadores hogareños contra incendio.

Para saber más sobre cómo reducir el riesgo de muerte visite el sitio www.usfa.fema.gov/50Plus.

Ella también perdió su casa.

Toda una vida de recuerdos puede hacerse cenizas en cuestión de segundos. Si usted tiene 50 años o más, preste atención en particular a la seguridad contra incendios. Por su bien y el de sus seres queridos.

- No fume cuando tenga sueño.
- Mantenga los calefactores por lo menos a un metro de las materiales inflamables.
- Nunca use ropa suelta mientras cocina.
- Revise las alarmas de humo todos los meses.
- Desarolle y practique un plan de escape en caso de incendio.
- Si existe la posibilidad, instale rociadores hogareños contra incendio.

Para saber más sobre cómo reducir el riesgo de muerte visite el sitio www.usfa.fema.gov/50Plus.

A Fire Safety Campaign for People 50-Plus

Información De Seguridad Contra Incendios Para Mayores De 50 Años

Cada año, approximadamente 1,100 americanos de más de 65 años mueren como consecuencia de un incendio hogareño. Comparadas con el resto de la población de Estados Unidos:

- Las personas entre 65 y 74 años tienen case DOS VECES más probabilidades de morir en un incendio.
- Las personas entre 75 y 84 años tienen case CUATRO VECES más probabilidades de morir en un incendio.
- Las personas de más de 85 años tienen más de CINCO VECES más probabilidades de morir en un incendio.

A través de una serie de acciones muy simples, las personas mayores pueden reducir en forma drástica el riesgo de morir o de resultar heridas en un incendio. Esto, sumado al dato de que las personas de más de 50 años forman parte de un grupo de alto riesgo, llevó a la Administración de Incendios de Estados Unidos (USFA, según la sigla en inglés), una división de la Agencia Federal para el Nacional Manejo de Emergencias (FEMA, según la sigla en inglés) y parte integrante del Departamento de Seguridad de Estados Unidos a desarrollar una campaña de seguridad a nivel nacional para los adultos de 50 años o más, sus familias y quienes los asisten. USFA le pide:

Evite Incendios. Salve Vidas.

Por su bien y el de sus seres queridos:
Implemente prácticas seguras para evitar incendios cuando fume, cocine y utilice calefactores. Mantenga las alarmas de humo en buenas condiciones, desarolle y practique un plan de escape en caso de incendio y, si es posible, instale rociadores contra incendio en su casa.

Fume en condiciones seguras

Sentarse en la silla favorita para fumar un cigarrillo después de cenar es para algunas personas un método fantástico para relajarse, pero el cigarrillo y la relajación pueden formar una combinación fatal.

Quedarse dormido mientras se fuma puede hacer que se encienda la ropa, las alfombras y el tapizado de los muebles. El consumo de alcohol y de medicamentos que provocan somnolencia agrava este riesgo.

Fumar en forma negligente es la principal causa de las muertes por incendio y la segunda causa de lesiones entre las personas de más de 65 años. Cuando no se los apaga bien, los cigarrillos siguen quemándose. Un cigarrillo en reposo que cae accidentalmente puede mantenerse encendido durante horas antes de iniciar un incendio. Antes de encender su próximo cigarrillo, recuerde:

- Nunca fume en la cama.
- Apague el cigarro o el cigarrillo ante la primera señal de somnolencia mientras mira televisión o lee.
- Utilice ceniceros hondos y moje las cenizas antes de arrojarlas a la basura.
- Nunca deje solos los cigarrillos encendidos y otros materiales para fumar.

Cocine con seguridad

Muchas familias se reúnen en la cocina para pasar un rato juntos, pero éste puede ser uno de los ambientes más peligrosos de la casa si no se adoptan prácticas seguras. Cocinar es la tercera causa de muertes por incendio y la principal causa de lesiones entre personas de más de 65 años.

Usar ropas sueltas, en especial mangas que cuelgan, dejar sola la cacerola en la hornalla o dejar materiales inflamables como agarraderas o papel de cocina cerca del fuego puede causar lesiones graves o inclusive la muerte. Ya sea que esté cocinando una cena especial para la familia o una comida frugal para los nietos:

- Nunca deje los elementos de cocción solos. Un incendio puede desatarse en cuestión de segundos.
- Use mangas cortas o ajustadas cuando cocine. Mantenga los repasadores, las agarraderas y las cortinas lejos del fuego.
- Nunca utilice las hornallas o el horno para calentar la casa.
- Revise dos veces la cocina antes de irse a dormir o de salir de su casa.

Calefaccione su casa con seguridad

En los meses de invierno, diciembre, enero y febrero, se producen más incendios hogareños que en cualquier otra época del año. Los aparatos de calefacción como las estufas y los hogares de leña hacen un ambiente confortable pero deben utilizarse con sumo cuidado. La calefacción es la segunda causa de muertes por incendio y la tercera causa de lesiones entre personas mayores de 65 años.

Muchas de estas muertes y lesiones podrían evitarse implementando prácticas de calefacción seguras. Antes de elegir un buen libro y de acomodarse frente al hogar, asegúrese de realizar las siguientes acciones:

- Mantenga las brasas dentro del hogar utilizando una parrilla lo suficientemente grande como para impedir que salga volando una chispa o que salga rodando un leño.
- Los calefactores ambientales necesitan espacio a su alrededor. Mantenga los materiales inflamables a una distancia de por lo menos un metro.
- Cuando compre un calefactor ambiental, busque uno que incluya la característica de corte automático de energía en caso de vuelco.

Plítica de seguridad contra incendios hogareños

- **Alarmas de humo**: Instale una alarma de humo en cada piso de su casa, verifique las baterías una vez por mes y cámbielas por lo menos una vez por año.
- **Plan de escape en caso de incendio hogareño**: Elabore y practique un plan de escape en caso de incendio en forma habitual, por lo menos dos veces por año. Mantenga las vías de salida libres.
- **Rociadores hogareños contra incendio**: Si existe la posibilidad, instale rociadores residenciales en su casa.

Para más información de prevención de incendios, por favor contáctese con:
Oficina de Publicaciones
Administración de Incendios de Estados Unidos
16825 South Seton Avenue
Emmitsburg, MD 21727
1-800-561-3356
www.usfa.fema.gov/50Plus

Sample Media Materials

Sample Media Cover Letter

Place cover letter to the media on your department's letterhead. Substitute appropriate information for the words in all-capital letters.

(INSERT DATE)

Dear (NAME OF MEDIA CONTACT):

Older people in (INSERT YOUR TOWN/CITY/DISTRICT) are more likely to die in a residential fire than the rest of our population. We are writing to ask (INSERT NAME OF MEDIA OUTLET) to support *A Fire Safety Campaign for People 50-Plus*. One important way that you can help is to use the enclosed (PRINT AD / LIVE READ PSA), which will provide older members of your audience, their families and caregivers with life-saving tips.

The majority of the approximately 1,100 fire deaths among Americans ages 65 and older each year are caused by careless smoking, cooking and heating practices. The campaign messages:

* Provide simple steps that people 50-plus can take to prevent death and injury caused by careless smoking, cooking and heating.
* Stress the importance of maintaining smoke alarms, developing and practicing a fire escape plan, and if at all possible, installing home fire sprinklers.

By using the (PRINT AD / LIVE READ PSA) you will help older members of our community avoid becoming a tragic fire statistic, and encourage people ages 50 and older who often care for and will soon enter this high risk group to develop strong fire safety and prevention habits before the risk escalates.

Enclosed you will also find a fact sheet about the campaign. I will contact you in the next couple of days to set up a meeting or phone call to discuss our request and answer your questions. Please feel welcome to call me at (XXX) XXX-XXXX. (FOR RADIO INCLUDE: WE WOULD ALSO BE GLAD TO RECORD THE PSA FOR AIRING ON YOUR STATION.) We will be deeply grateful to you for joining in our fire department's efforts on behalf of vulnerable older people in our community.

Sincerely,

NAME OF CHIEF, PIO, OR CAMPAIGN SPOKESPERSON

P.S. For more information about *A Fire Safety Campaign for People 50-Plus*, please visit the United States Fire Administration's campaign Web site at www.usfa.fema.gov/50Plus.

Sample News Release

Place news release on your department's letterhead. Substitute appropriate information for words in all-capital letters. To include data from your state in paragraph #4 of the sample release, go to the "data report" with home fire fatality statistics on people ages 50 and older on the campaign computer CD or www.usfa.fema.gov/50Plus.

Contact: INSERT YOUR NAME
YOUR TELEPHONE NUMBER

(YOUR FIRE DEPARTMENT) PARTNERS WITH USFA TO KICK OFF NEW FIRE SAFETY CAMPAIGN FOR PEOPLE AGES 50 AND OLDER

Community-wide Effort to Lower Escalating Fire Death Rates Among Older Population from Careless Smoking, Cooking and Heating

(CITY), (DATE) – The (CITY, TOWN) Fire Department today launched *A Fire Safety Campaign for People 50-Plus* to prevent the community's older citizens from joining fire death statistics that escalate dramatically beginning at age 65. The fire department is conducting the campaign in partnership with the Department of Homeland Security's Federal Emergency Management Agency (FEMA).

Each year, approximately 1,100 Americans ages 65 and older die in home fires and another 3,000 are injured. People ages 65 and older are three times as likely to die in a residential fire as the rest of the population. Careless smoking, cooking and heating practices are the top three causes of fire deaths among people ages 65 and older.

"Our firefighters will teach people simple steps for safe smoking, cooking and heating their homes – steps that will help them avoid unspeakable tragedy," said (INSERT CHIEF'S NAME AND TITLE). "We will also encourage our older citizens to maintain their smoke alarms, develop and practice a fire escape plan, and if at all possible, install home fire sprinklers." Chief (XXXXXX) noted that the campaign reaches out to the 65-plus population and to people ages 50 and older who often care for and will soon enter this high risk group.

In (YOUR STATE) people ages 65 and older were (XX) as likely as the rest of the population to die in a residential fire from 1989 through 1998. In that decade (XXX) adults ages 65 and older died in residential fires, according to the U.S. Fire Administration (USFA).

"We welcome invitations from community and faith-based organizations to present fire safety programs for people ages 50 and older, their families and caregivers," said Chief (INSERT CHIEF'S LAST NAME). "Our firefighters will also make home visits to point out fire hazards, help plan escape routes and provide fire safety materials."

Contact the fire department at (INSERT CONTACT NUMBER) to schedule a fire safety presentation or request a home visit. *A Fire Safety Campaign for People 50-Plus* campaign materials are available through the fire department or via the Web at www.usfa.fema.gov/50Plus.

#

Sample Radio Public Service Announcements (PSAs)

Live-Read Scripts

FIRE SAFETY FOR PEOPLE 50-PLUS (:30)

DID YOU KNOW THAT PEOPLE AGES 65 AND OLDER ARE THREE TIMES AS LIKELY TO DIE FROM FIRE AS THE REST OF US? CARELESS SMOKING, HEATING AND COOKING ARE THE LEADING CAUSES OF FIRE FATALITIES FOR AROUND 1,100 OLDER AMERICANS EVERY YEAR. IF YOU ARE OVER THE AGE OF 50, BEGIN NOW TO PRACTICE FIRE-SAFE HABITS IN YOUR HOME. FOR FIRE SAFETY TIPS TO IMPROVE YOUR WELL-BEING AND PEOPLE IN YOUR CARE, CONTACT **(INSERT NAME OF FIRE DEPARTMENT)**.

FIRE SAFETY FOR PEOPLE 50-PLUS (:60)

AROUND 1,100 PEOPLE AGES 65 AND OLDER WILL DIE FROM FIRE THIS YEAR. THAT'S THE AVERAGE…ALMOST 3 PEOPLE A DAY. CARELESS SMOKING, HEATING AND COOKING ARE THE LEADING CAUSES OF FIRE DEATHS AND INJURIES AMONG OLDER AMERICANS. IF YOU ARE 50 OR OLDER, BEGIN <u>NOW</u> TO PRACTICE FIRE-SAFE HABITS IN YOUR HOME. IF YOU SMOKE, PUT YOUR CIGARETTE OUT – ALL THE WAY OUT – AND DON'T SMOKE WHEN YOU'RE SLEEPY. AVOID LOOSE CLOTHING WHEN YOU COOK – AND NEVER WALK AWAY FROM THE STOVE WHEN FOOD IS COOKING. USE A FIREPLACE SCREEN – AND KEEP FLAMMABLE MATERIALS AT LEAST 3 FEET AWAY FROM HEATERS. TO MAKE SURE YOU GET OUT ALIVE IF FIRE DOES HAPPEN, MAINTAIN YOUR SMOKE ALARMS. PRACTICE A FIRE ESCAPE PLAN. AND, IF POSSIBLE, INSTALL HOME FIRE SPRINKLERS. FOR MORE FIRE SAFETY TIPS TO IMPROVE YOUR WELL-BEING AND PEOPLE IN YOUR CARE, CONTACT **(INSERT NAME OF FIRE DEPARTMENT)**.

Sample E-Mail Message

Please include contact information for your fire department and send e-mails to friends and supporters. Encourage them to forward the e-mail to friends and relatives.

Dear Friend:

Each year, approximately 1,100 Americans ages 65 and older die as a result of a residential fire. That is nearly three people each day. If you or someone you love is over the age of 50, begin NOW to practice fire safety in your home.

Careless smoking, heating and cooking are the leading causes of fire fatalities among older people. With a few simple steps you can protect yourself and those in your care from unspeakable tragedy. Never smoke when you're drowsy and always put cigarettes out in deep ashtrays. Avoid loose clothing when you cook and never walk away from the stove when food is cooking. Keep clutter and flammable materials at least three feet away from heaters – and use a fireplace screen. Remember to maintain smoke alarms, practice a fire escape plan and, if possible, install home fire sprinklers.

Please forward this important fire safety message to your friends and relatives. It is never too early to develop strong fire safety habits.

For more information on *A Fire Safety Campaign for People 50-Plus*, please contact our fire department at (XXX-XXX-XXXX) or visit the U.S. Fire Administration's Web site at www.usfa.fema.gov/50Plus.

Sincerely,

(INSERT YOUR NAME AND CONTACT INFORMATION)

Please Send Us Your Feedback!

A Fire Safety Campaign for People 50-Plus is a new campaign developed by the United States Fire Administration for our Fire Service partners. Working together, we can help vulnerable older people reduce their risk of death and injury from fire and assist their families and caregivers with fire safety tips. This *Campaign Guide* includes campaign materials that can also be found on the campaign computer CD and the campaign Web site – www.usfa.fema.gov/50Plus. The materials are free and can be copied as many times as you would like.

We would love to hear from you. Our materials are produced to help you reach out into the community through presentations and home visits, and work with the media. Please provide us with the following information, including any suggestions or comments about *A Fire Safety Campaign for People 50-Plus*.

Name   Position

Fire Department

E-Mail Address

Street Address

City   State   Zip Code

(   )   (   )
Telephone number   Fax number

Suggestions and Comments:

Return form by fax to: 301-447-1102

Return form by mail to: Kathy Gerstner, United States Fire Administration, 16825 South Seton Avenue, Emmitsburg, MD 21727

Thanks so much for your feedback! We will use it to improve and enhance the campaign.

For additional fire prevention and safety information, please contact:

Publications Office
United States Fire Administration
16825 South Seton Avenue
Emmitsburg, MD 21727
1-800-561-3356
www.usfa.fema.gov/50Plus

www.ingramcontent.com/pod-product-compliance
Lightning Source LLC
Chambersburg PA
CBHW081241170526
45165CB00009B/3142

9 781482 662054